1

2

3

4

5

6

7

8

9

10

For my Mum and Dad

First published in 1986 by
Deans International Publishing
52–54 Southwark Street, London SE1 1UA
A division of The Hamlyn Publishing Group Limited
London · New York · Sydney · Toronto

Text and illustrations Copyright © Deans International Publishing,
a division of The Hamlyn Publishing Group Limited, 1986

ISBN 0 603 00717 1

Printed and bound by Purnell Book Production Ltd.,
Paulton, Bristol.
Member of BPCC plc

Harry Hippo's Counting Games

Written and illustrated by Gillian Chapman

Dean

Harry Hippo was usually a very happy hippo. He had fun and games with everything, helped his Mum around the house, and had lots of animal friends.

But today was different; young Harry was troubled. He was finding it very difficult to learn to count to ten.

Harry's Mum said she would teach him
as soon as she had finished her cleaning.

"I can't wait," he told himself.
"I'll go and ask my friends to help.
It's market day in Jungle Town.
Lots of my friends will be around."

Mr Sprout the Greengrocer was busy
serving a customer. Harry asked him
if he would help him to count.

"Of course," said Mr Sprout.
"Take Miss Antelope's shopping list.
It goes like this . . .

"**One** crisp green cabbage,
two juicy grapefruit, large and yellow,
three ripe golden pears,
four green apples and
five bent bananas in a bunch.
Six soft tomatoes – going cheap,
seven crunchy carrots,
eight fresh onions on a string,
nine plump purple plums and
ten muddy potatoes."

"Delicious!" said Harry.

Miss Antelope has finished her shopping
and hurries home to start her cooking.
Oh, she's dropped **one** cabbage
and **two** carrots in the road!

Harry met Ben Rhino but he wasn't interested in Harry's game, as he could already count to ten. Harry wanted to show what he'd learned.

"I've got **two** buttons on the front of my trousers and you've got **three** buttons on your shirt," he said to Ben.

Next Harry saw Miss Meringue.

"I hear you are learning to count,"
she said. "Let's count my delicious cakes
and bread, that will help you to remember.
I've got **one** sponge — all chocolaty,
two fruit cakes with nuts on top,
three apple pies,
four jam tarts, round and sticky,
and **five** long crusty loaves.

"**Six** chocolate eclairs, filled with cream,
seven iced currant buns,
eight doughnuts sprinkled with sugar,
nine cherry cakes
and **ten** soft rolls freshly baked today."

Harry felt hungry.

Harry spied the greedy Tiger Twins munching **four** jam tarts that were meant for their tea.

Harry felt very, very hungry.

Poor Harry's tummy started to rumble and kind Miss Pandora guessed why, so she made up a little counting game.

"You can have one if you count for me. How many cakes have a bright red cherry?"

"There's **five**!" Harry cried, without delay, popping one in his mouth straight away.

Pandora clapped her paws and laughed. "That's good Harry, you're learning fast!"

Feeling very pleased, Harry went over
to Mr Creak the Handyman's stall.

"I'll help you, Harry, if I can,"
he said kindly.
"Here's **one** stepladder,
two brooms with long handles,
three doormats,
four plastic watering cans and
five scrubbing brushes.

"**Six** china teapots,
seven striped umbrellas,
eight balls of strong string,
nine different sizes of saucepans,
and **ten** clothes pegs in a bucket."

"Goodness me!" thought Harry.
"What a muddle!"

Then Harry met poor Mr Toddle, looking very miserable. He'd just dropped all his shopping.

"I've lost **six** lids. How my wife will moan. She'll see they are missing when I get home."

So Harry helped him to look for the lids.

After all that searching in the grass,
Harry sat down for a minute to have a rest.

Mrs Sparkle and her family bustle by.
"Hurry up at the back," Harry hears her cry.
There's no time to chat, we mustn't delay.
With **seven** children, I'm busy all day."

Harry counted them as they scurried after
their Mum.

"Hello Harry," chirped Pete the Painter.
"I'm smartening up this house of mine.
I'll sing you a colourful counting rhyme.

"Here's **one** tin of red to paint the door.
Careful! Don't spill it on the floor.
Two tins of orange and **three** of green,
the brightest colours you've ever seen.
Four tins of yellow and **five** of blue,
to paint the woodwork good as new.

"For the roof and chimney there's **six** tins of black
and **seven** of white for the front walls and the back.
I've **eight** paint brushes and **nine** mates to help.
It's too big a job to do by myself.
Ah! here's the tea. That's what we all need.
And **ten** homemade biscuits sprinkled with seeds."

Harry loved Pete's song and was dazzled
by the colours.

Harry walked on.
It was a warm afternoon and in
the shade the Ice-cream Lion was
selling lots of ice-creams. "Can you count
eight ice-cream cones, Harry?" he called.

Nearby, Almond the Camel was feeling very cross.

"That chimp is juggling with **nine** of my best coconuts," he yelled.

Harry watched in amazement.

"Here's some peanuts for you, Harry. You're a good boy," said Almond.

Babs the Balloon Seller dozed in her deckchair as Harry walked by. She smiled at him.

"I haven't many balloons left to sell. I've been busy today and done very well. There's **eight** in my bunch and the Tigers have **two**. If you can count **ten**, I'll give one to you."

Harry was so happy. He counted **ten** balloons and had a lovely red one as a prize.

Harry wandered over to the waterhole.

"I don't need my friends' help. I can count now by myself. I can see **one** canoe on the lake, with the **two** Tigers in it. There are **three** monkeys playing hide-and-seek, with **four** snakes looking for them. **Five** parrots are picnicking in that tree.

"That lion cub is just about to eat **six** iced buns, and **seven** ants are crawling up his leg – Ha! **Eight** noisy ducks are squabbling in the water, **nine** bees are buzzing round my head and I've got **ten** peanuts left to take home to Mum."

Which reminded Harry that it was about time he set off for home.

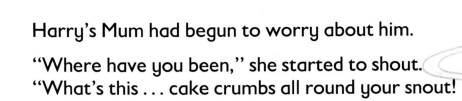

Harry's Mum had begun to worry about him.

"Where have you been," she started to shout.
"What's this . . . cake crumbs all round your snout!

"Where's your counting frame, Harry?" said she.
"I'll help with your numbers after tea."

Harry smiled to himself secretly.
He knew how surprised his Mum would be!

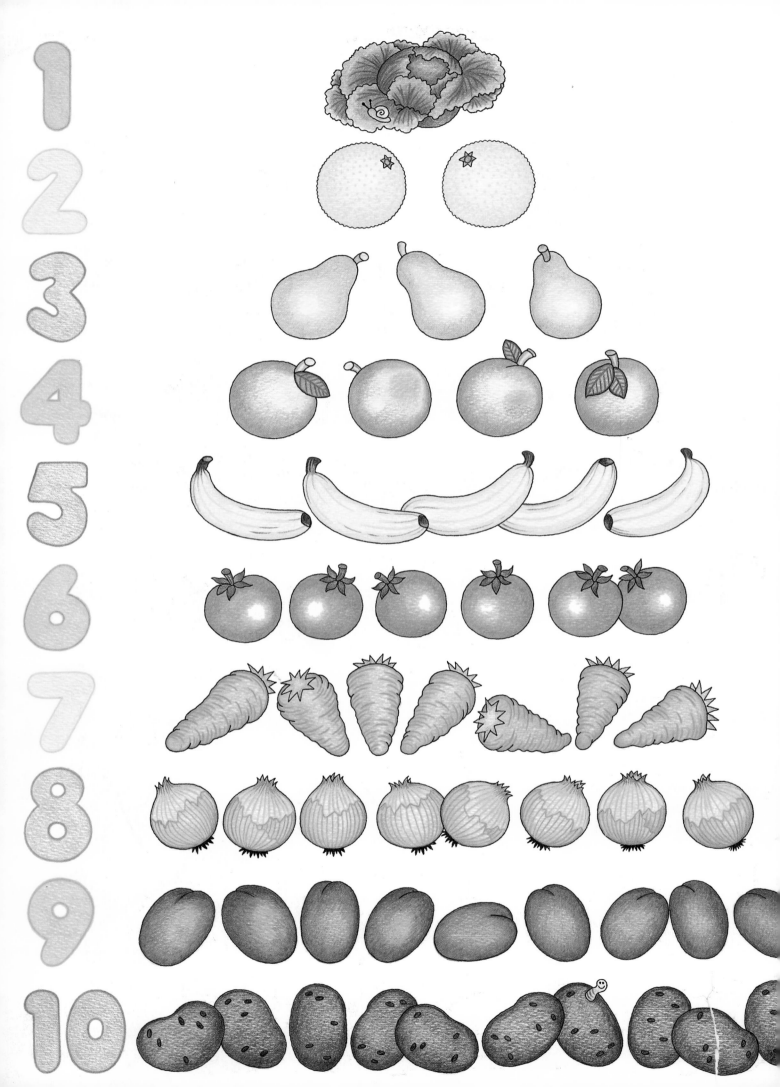